EDUCATIONAL CONTINUITY DURING UNCERTAINTY

Online Learning

Considerations for Educators

To request permission, contact the publisher at info@lctelearning.com.

www.lctelearning.com

CONTENTS

Educational Continuity During Uncertainty

ABOUT..1

INTRODUCTION..2

SOCIAL CONTINUITY......................................4

FAMILIAL CONTINUITY..................................16

EMOTIONAL CONTINUITY.............................26

COGNITIVE CONTINUITY..............................35

ACADEMIC CONTINUITY...............................43

BIBLIOGRAPHY...51

ABOUT

This book offers valuable research-based strategies for teachers to tackle remote, distance, hybrid, or blended learning. It starts by defining these five key principles educators can implement to achieve continuity with stakeholders.

1. Social
2. Familial
3. Emotional
4. Cognitive
5. Academic

Next, it provides teaching and learning activities for each of the five continuity principles which can be applied to synchronous and asynchronous experiences.

Finally, various technological tools are suggested for these activities. It is important to note that the technological tools are being updated all the time. New applications that will do the same job may even be developed. Teachers should stay abreast of these updates to make sure they are using the most up-to-date and appropriate tools.

INTRODUCTION

Communication, collaboration, connection, and content delivery with students look different during online learning. This new normal for educators, students, and families bring about much uncertainty.

Educators can draw upon their creativity, innovation, flexibility, and grit to help during these tough times –much like they have always done.

Without a doubt, educators will face challenges. They do, however, have the strength, ability, and resilience to equip themselves when re-designing learning activities and rethinking the ways students engage in their online classes.

In addition, educators can help soften the negative academic, social, and emotional effects of online learning on students. They can even support parents and families with consistent communication. This can be done when educators see how they can implement activities that foster relationship building, engagement, social and emotional learning, higher-order thinking skills (HOTS), and metacognition.

Ultimately, educators can use the following five considerations to guide them in achieving continuity with their students, parents, and families:

1. Social Continuity
2. Familial Continuity
3. Emotional Continuity
4. Cognitive Continuity
5. Academic Continuity

SOCIAL CONTINUITY

Social continuity relates to the relationship and trust-building that takes place between students and their peers and teachers. Small and whole-group activities, as well as collaborative instructional methods during synchronous and asynchronous learning experiences, help with social continuity.

Teachers can achieve positive student-to-student and teacher-to-student interactions through virtual, collaborative activities. They can do this with virtual icebreakers and circle time. Teachers can also build respectful relationships, routines, and responsibilities with students through engaging and interactive virtual scavenger hunts and field trips.

Some activities to consider include:

1. Virtual Icebreakers & Celebrations
2. Virtual Scavenger Hunts & Fliphunts
3. Virtual Field Trips & Guest Speakers
4. Virtual Circle Time & Story Telling
5. Digital Breakouts/Escape Rooms

Teachers must consider whether the technological tools they are going to use for these activities are suitable. For example, they can use Microsoft OneNote for digital breakouts.

Teachers should also think about using a single platform like Microsoft Teams, Google Classroom, Blackboard, Canvas, or other learning management systems that have features suitable for collaboration and communication activities that give social continuity.

These features include but are not limited to

- an area for whole/large group discussions
- a small group discussion/conversation spot
- breakout rooms during online class meetings
- a tool to host live, online class meetings
- an assignment tool
- a grade book
- a space to have private chat
- a place for announcements
- integration with other tools like cloud storage
- area to centralize learning materials

Virtual Icebreakers & Celebrations

Teachers can use the channels option in Microsoft Teams or create Flipgrid topics for icebreaker questions and celebrations. Here are some examples of icebreakers teachers could use for online learning:

1. The teacher puts up a passage and asks students to connect to it with one word or phrase
2. Students find a place on a map and describe why they want to go to this place or talk about the location if they have visited or lived there before
3. Students locate an object in their surroundings and use it to symbolize some aspect of themselves or their lives
4. Students share a goal they have achieved or would like to achieve for the day, week, month, etc.
5. The teacher reads a quote and asks students to relate to it or connect it to what they are learning (e.g., a book they are reading, a historical figure they are studying)

To accomplish the icebreaker tasks, teachers can create an "Icebreaker and Celebrations" channel in their Microsoft Teams class team. Here, they provide a place for students to respond and reply to each other in an online post.

See an example of an icebreaker channel in Microsoft Teams.

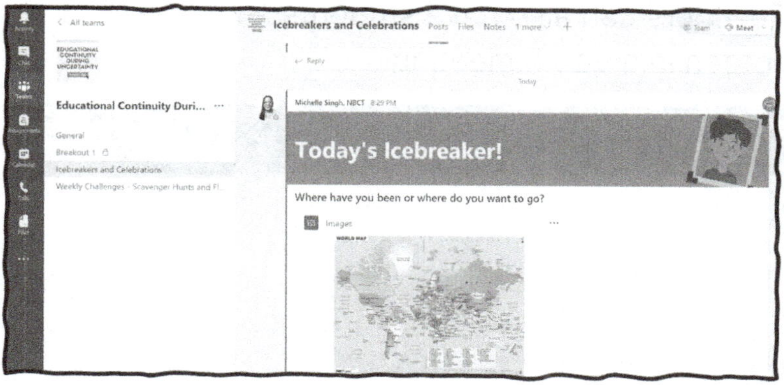

Teachers can also create a Flipgrid topic and share this with students to allow them to record video or audio responses. See an example of a Flipgrid Icebreaker.

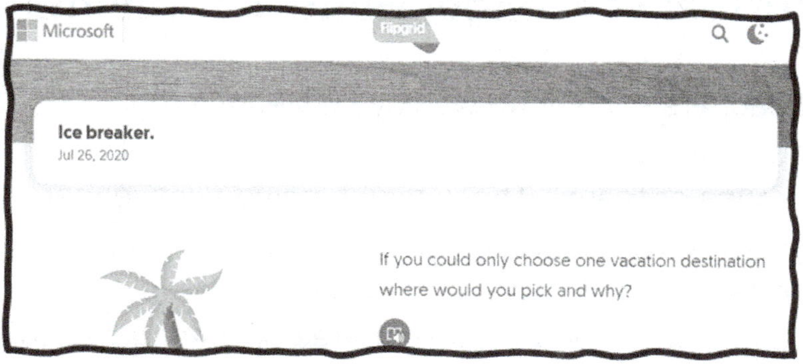

Finally, during a live lesson or online class meeting, teachers can ask students to answer questions within the meeting chat, to write responses on the meeting whiteboard (this is an integration with Microsoft Teams), or share/speak their responses out loud.

See an example of the whiteboard integration being used in a Microsoft Teams meeting. Teachers can also have students annotate on a screen in a Zoom meeting.

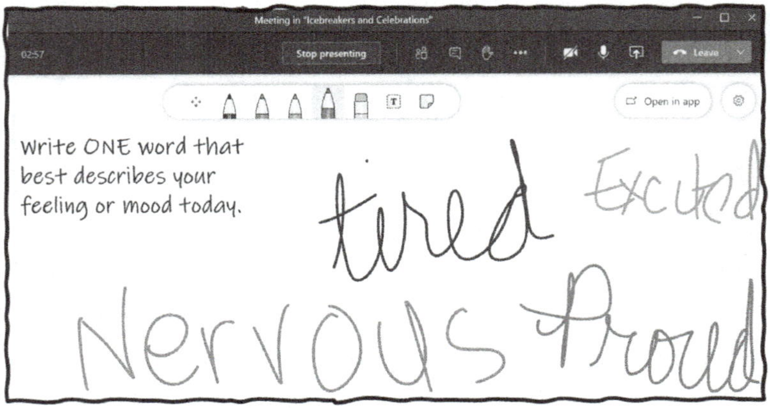

Virtual Scavenger Hunts & Fliphunts

Teachers can set up weekly challenges for students to complete individually or in small groups or teams. These might include scavenger hunts or Flipgrid Fliphunts. Fliphunts are engaging and exploratory learning tasks for students that can be used for any subject area or grade level. In a Fliphunt, students complete multiple tasks to solve problems. The tasks tap into various learning styles which give students voice and choice as they participate. Teachers can create Fliphunts on Flipgrid for students to complete.

See an example of a Fliphunt Challenge on Flipgrid.

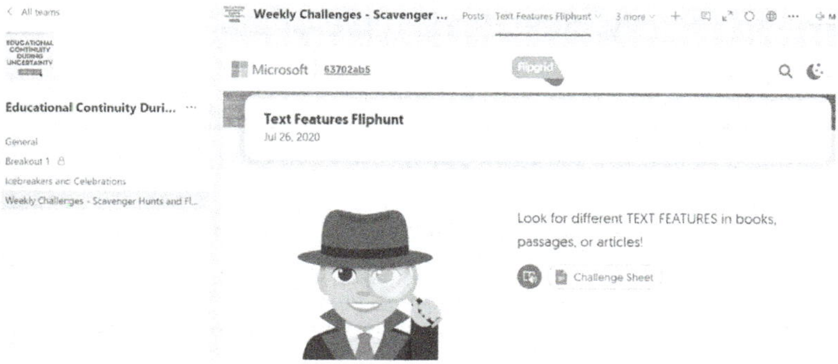

Students can work together in small groups in a private channel in Microsoft Teams to complete their Flipgrid Fliphunt or scavenger hunt activity. This can be done asynchronously as long as students have shared responsibilities for their learning tasks. This shared responsibility means students complete whatever part of the learning task they take on, they help with or contribute to the other responsibilities for the group, and review the completed group task before final submission.

Complete → Contribute → Review

In addition, during live lessons or online class meetings, students can work on a collaborative Word, Google, or Kami documents to complete a scavenger hunt activity about their class meeting topic.

This can also be done in breakout rooms where students are in small groups and are working synchronously. See an example of a Scavenger Hunt activity.

Literary Devices in The House on Mango Street

Group Scavenger Hunt

DIRECTIONS: Work in your small group breakouts to identify at least five literary devices from the vignettes we have read. Complete the table below. Be sure to have a discussion and come to a consensus on the meaning of each device. When you return to the whole group in the class meeting, be prepared to share two of your findings with the class.

Vignette	Page #	Direct Quotation	Literary Device	Discussion/Meaning

Virtual Field Trips & Guest Speakers

Teachers can use Flipgrid Virtual Field Trips (formerly Skype in the Classroom) to schedule virtual field trips and guest speaker experiences for their students. Flipgrid Virtual Field Trips (formerly Skype in the Classroom) is a free resource and a global community that can be used by educators and experts to connect.

These experiences can be found on the Flipgrid platform. Flipgrid is a discussion platform. Educators and experts create video topics with the aim of engaging students in global learning experiences. Flipgrid has a Discovery Library with thousands of topics teachers can use for virtual field trips and guest speakers. Teachers can also visit the Microsoft Events page where they can sign up for live events with guest speakers.

Teachers and students can easily connect with authors and experts worldwide. Experts come from museums, historical sites, national parks, and more. Teachers can take their students on a live virtual field trip with authors and experts by registering on the Microsoft Events registration page. Teachers can also take their students on virtual field trips by viewing pre-recorded videos in the Flipgrid Discovery Library. Teachers can bring these experiences to students synchronously or asynchronously. Students watch the video or view the live event. Students then answer questions about what they have learned in a discussion post or via a Flipgrid topic. Students can even prepare questions to ask the expert during these learning experiences.

In a nutshell, teachers are able to bring global experiences to students. With a single meeting, teachers can break down barriers and take students outside the walls of their home or classrooms so they can connect with other cultures and countries, experts, and classrooms. Student learning experiences can be enriched if teachers take advantage of these free resources provided by Flipgrid Virtual Field Trips.

Virtual Circle Time & Story Telling

Teachers can have circle time activities during a live online class meeting, sharing a story or something about a topic with their students. Teachers can also ask their students to share a story, add to the story, or share something about a topic. Another option for teachers during a live online class meeting is to use breakout rooms for pairs or small groups of students. Students can meet in breakouts to share and respond to each other. See examples of circle time activities from The Responsive Classroom website, *The Morning Meeting Book*.

Around-the-Circle Sharing - all members of the circle say something very brief about the story or topic.

Partner or Small-Group Sharing - students pair up to talk with each other about the story or topic.

Dialogue Sharing - one student presents something about the story or topic and then asks for responses from the class.

If teachers want to do this so their students can participate asynchronously, they can record a Flipgrid video of themselves reading a story or sharing information about a topic. They can ask students to watch the video and respond in their own Flipgrid video responses. The URL or link to this Flipgrid activity can be shared with students as an announcement or assignment in any Learning Management System or communication platform.

Digital Breakouts / Escape Rooms

A digital breakout allows students to review the content provided by the teacher about the topic. Then the student must solve a puzzle, problem, or task to move on. Students can solve a series of topics to get to the end of the breakout where there is some sort of incentive or prize. Examples include a digital breakout about the chapters in a book or textbook or a general topic with interdisciplinary connections. Teachers can use Flipgrid or Microsoft OneNote to create these problem-solving tasks for students to work on in small groups.

Teachers can create password-protected sections in Microsoft OneNote with the content students need to review and learn to problem-solve. The passwords for these sections are the answers to a question about the content a student has learned. When the correct password is entered, students unlock the next section and can complete the tasks or solve the problems in that section. Students do this until they get to the final topic and the last section of the Microsoft OneNote digital breakout. In the same way, teachers can create Flipgrid topics and add passwords for students to access them.

See the examples below of Microsoft OneNote and Flipgrid digital escape rooms or breakouts.

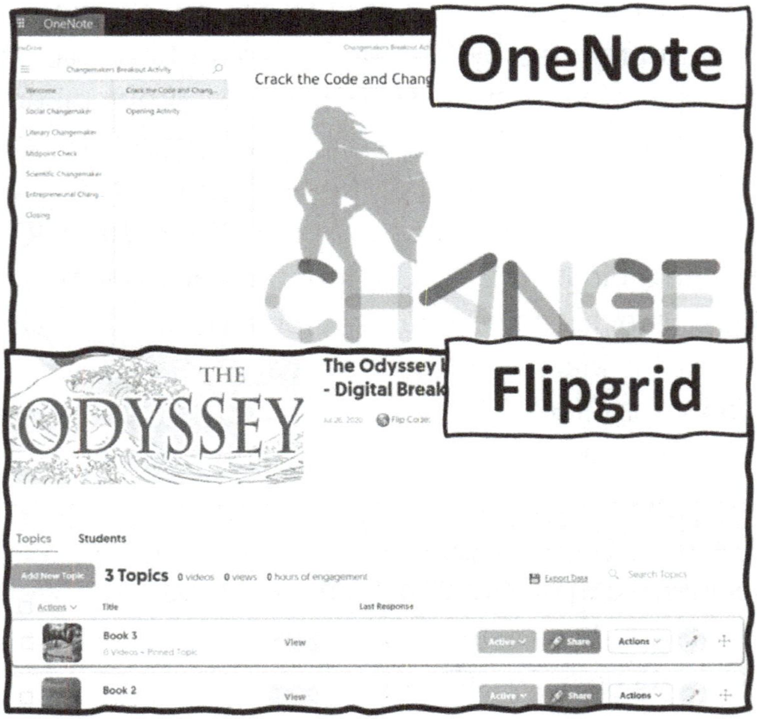

To implement digital breakout activities with their students, teachers can post the URL or link to their Microsoft OneNote or Flipgrid in their learning platform as an assignment, announcement, or post.

While in a live online class meeting, students can work synchronously in small groups in breakout rooms to complete the activity. Also, students can work in private channels set up by the teacher in Microsoft Teams. They can use the "Meet Now" feature to work on the Microsoft OneNote breakout and record their Flipgrid video responses for their Flipgrid Breakout.

For the asynchronous option, teachers can post the digital breakout URL for students to complete individually. Students can also work in groups asynchronously. In a private channel in Microsoft Teams, small groups of students can use the post feature to communicate with each other about tasks to complete and set deadlines to complete the breakouts.

Get Started Now!

1. Decide on your main topic or subject
2. Choose your subtopics (these will be the sections in your Microsoft OneNote or your Flipgrid topics)
3. Set up your Microsoft OneNote by adding a Welcome section with rules, guidelines, and resources for students
4. Add your password protected sections for students to break out of or create your Flipgrid topics
5. Make your last password protected section or topic a celebratory one that congratulates students on completing the breakout

FAMILIAL CONTINUITY

Familial continuity happens when teachers provide increased opportunities that will engage parents and families in the learning of their children. Now, more than ever, parents and families are an integral part of their children's educational setting and experiences. During online learning, the expectation is that students will learn at home with guidance and resources from their teachers and support from their parents and families. There are many communication tools that can be used with parents and families (e.g., Remind and Class Dojo). The challenge is that there is a lot of information teachers share with parents, and this content comes in multiple formats. For example, a teacher may share a survey, video, document, picture/graphic, appointment, etc. When these are opened by the parent, they may open in different applications which can be confusing.

In addition, the organization of this information, when shared with parents and families, can get chaotic for them to keep track of and refer to later if they need to.

This can be even more confusing for parents and families when they have multiple students in the household who are using different apps and tools. To streamline all of this for parents and families, teachers should consider using Microsoft OneNote or Wakelet Spaces.

Microsoft OneNote is a digital notebook with sections that are similar to dividers in a traditional notebook. Inside each section of the notebook are pages. Teachers can add a variety of content on these pages including typed text, hyperlinked text, images, audio recordings, embedded videos, printouts of PDFs, Word documents, Excel spreadsheets, PowerPoint presentations, and more.

Wakelet is a platform that allows its users to save and organize various forms of content including videos, images, text, weblinks, etc. in one place. Wakelet Spaces is part of the Wakelet platform that creates a single page to house a variety of Wakelet collections for easy access with one link or code.

Teachers can create a Microsoft OneNote notebook and share a public, view-only link with parents and families. Teachers can use a URL shortener to make this long link short and simple for parents and families to have and refer to quickly. The link shared with parents does not change. This allows them to keep the same link all school year.

When the teacher creates something on Microsoft OneNote, they can add sections such as announcements, resources, tutorials, and much more. These sections can contain pages where parents can locate helpful content. The teacher organizes and updates the Microsoft OneNote weekly. When the parents and families access Microsoft OneNote, they can easily view the information in sections and navigate to what they need.

Also, applications such as a survey in Microsoft Forms or a YouTube tutorial video can easily be embedded into the Microsoft OneNote page keeping parents and families in one place. To take it even further, parents can use the Immersive Reader for language and reading support. See this example of OneNote for parents.

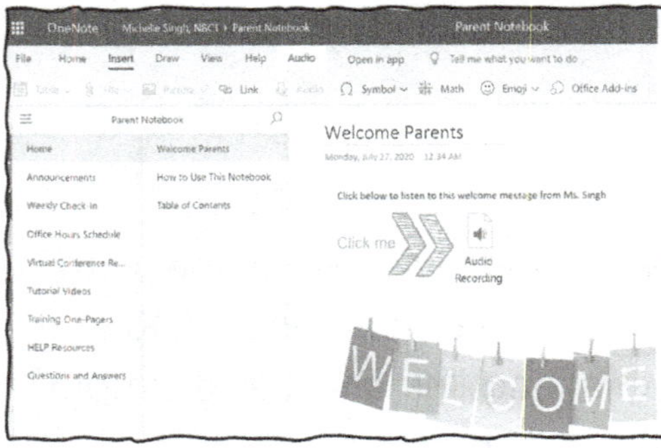

Include multi-modal content on the Microsoft OneNote pages. These include audio recordings, images, embedded videos, infographics, etc.

Teachers can also use Wakelet to curate all the content they have for parents in a Wakelet Space. This Wakelet Space is a single, clean space for collections of information like newsletters, tutorials, announcements, surveys, homework guides, videos, and other resources. Teachers can add the information they would like parents to access and invite them to access these resources via a URL, email, or QR code.

Before discussing examples for familial continuity, here are some basic considerations and actions teachers can take to help parents and families feel less overwhelmed and more encouraged during online learning.

- Streamline communication with parents by using **one platform**, so parents have a consistent place to look for information from teachers. Microsoft OneNote and Wakelet Spaces, previously discussed, are recommended. A teacher can also host their own website for parents and families.
- Share information, expectations, announcements, etc. **weekly** and not daily to prevent parents from being inundated with messages, especially if they have more than one child.

- Offer **"how-to"** resources to give guidance, training (especially with technology), and information to parents, so they can help their children complete learning activities. These can be one-pagers or videos.
- Explain the **"why"** by defining the big picture, so parents know the scope of what their children will learn and do. This can be a simplified version of the learning objectives/standards or "I can" statements which helps them understand expectations for their children.
- Be flexible in the way information is shared with parents and families. This means teachers should have **multi-modal** options when communicating with parents and families. Teachers can provide written, audio, and visual communication with parents. They can do this by using Nearpod or Microsoft Sway to create newsletters.

In terms of examples of what teachers can do to achieve familial continuity, the following can encourage engagement with parents and families:

- Create a check-in survey where parents and families can have a way to share and ask questions in private.
- Offer after-hours support by having a "bank" of resources parents and families can access anytime.

- Host virtual office hours and conferences where parents can have live meetings with the teacher (students can be included as well).
- Have a public question forum for parents and families to post and get feedback from the teacher and other parents and families.
- Share HELP resources to support students, parents, and families as they navigate the stressors of online learning.

Check-In Survey

Teachers can use Microsoft Forms, Google Forms, or another survey creation platform to create a check-in survey for parents and families. This will provide a chance for parents, families, and students to share successes and struggles and ask the teacher questions privately.

These questions can be used for the check-in survey:

1. Would you like to share any triumphs or successes that you and your child experienced this week?
2. Would you like to share any struggles or roadblocks you and your child experienced this week?
3. Do you have any questions you would like the teacher's help with answering?

After-Hours Support Videos & One-Pager Documents

Teachers can provide a "bank" of resources parents and families can access 24/7 in the form of short video tutorials, training, and one-pagers. Video links from YouTube, Flipgrid Shorts, Prezi Video, or WeVideo can easily be embedded or added into a Microsoft OneNote page. Links to videos can also be added to Wakelet. This makes it simple for teachers to add these resources and for parents to view them in one place. One-pager files (Word, PowerPoint, or PDF) can be inserted as printouts into Microsoft OneNote and can easily be added to Wakelet. When teachers create and give access to these support systems for students, parents, and families, it empowers them to get the help they need when they need it. See the example of the Microsoft OneNote Parent Notebook.

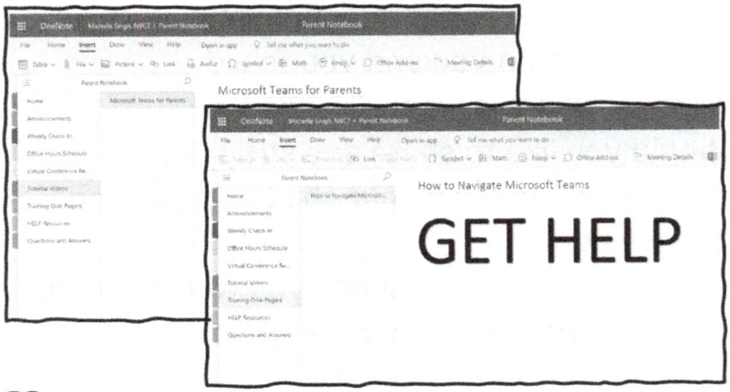

Virtual Office Hours & Conferences

Teachers can establish open office hours for parents once a week by scheduling times on their Outlook calendars using the Microsoft Teams or Zoom add-in option. They can copy and post the meeting invitation link in the Microsoft OneNote or add it to the Wakelet for parents to access and join at their convenience. See the example below of the Outlook calendar invitation details being added to the OneNote page. Further, teachers can directly schedule virtual conferences with individual parents and families. They can set up a Microsoft Teams or Zoom meeting and send email confirmations to parents.

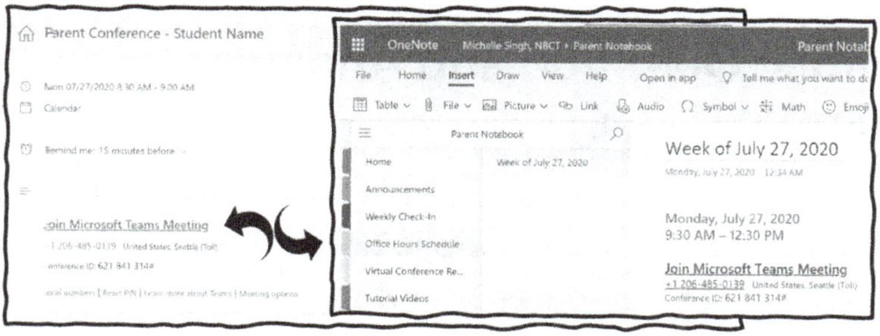

Teachers can use the Microsoft Bookings platform. Microsoft Bookings is a scheduling tool that is part of Microsoft 365. It syncs with the Outlook calendar of teachers who use it. It also sets up automatic meetings using Microsoft Teams. With Microsoft Bookings, parents can schedule meetings with teachers and choose times that are convenient and open both. When the parent meetings happen in Microsoft Teams, teachers and parents can use certain features for accessibility such as live captions and transcriptions.

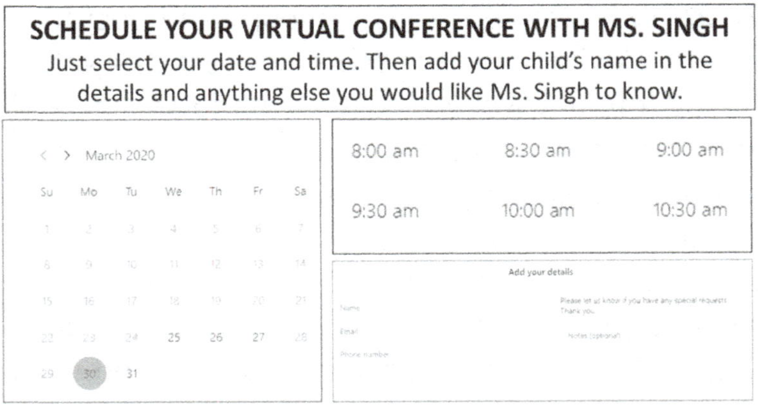

Public Question and Answer Forum Using Flipgrid

Teachers can set up a public question-and-answer forum for parents and families to post and get feedback from the teacher and other parents and family members. This can be easily accomplished using Flipgrid. Teachers set up a "Parent Q&A Corner" Flipgrid and add a guest password to share with parents and families. Copy and paste the Flip link and guest password in the Microsoft OneNote for parents to have easy access to view and add to the Q & A.

See the Microsoft OneNote example with the Flipgrid Parent Q&A Corner. Teachers can also add the Flipgrid to the Wakelet Space.

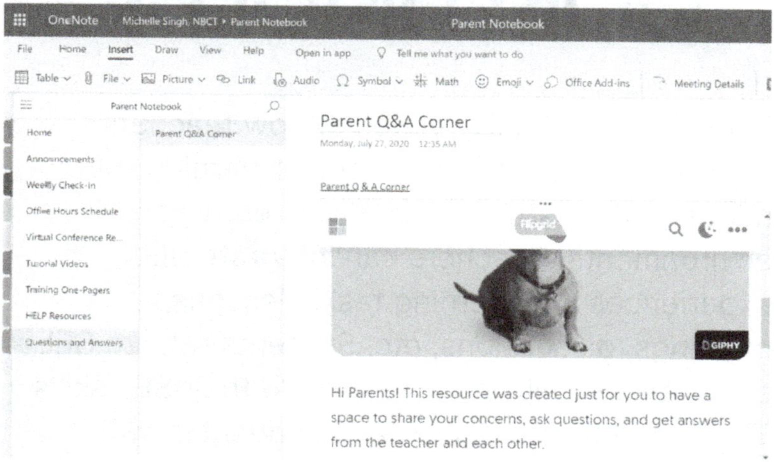

Help & Community Resources

Teachers can share resources they find beneficial for parents and families as they navigate the stressors of online learning. These resources can relate to mental health, stress, anxiety, time management, organization, and free community activities. Teachers can gather resources and add them to the "HELP Resources" section in the OneNote. Examples include links to the free virtual workshops, camps, training, events, etc. offered by Microsoft, colleges and universities, community organizations, non-profits, religious groups, and mentor groups of local fraternities and sororities.

EMOTIONAL CONTINUITY

Emotional continuity relates to how teachers address the social and emotional learning (SEL) needs of their students in online learning environments. Teachers can integrate SEL experiences into learning tasks, lessons, meetings, discussions, etc. Students require SEL experiences daily in order to build their SEL skills. They need to feel valued and connected with their learning environment, especially online where they do not have physical connections to their peers and teachers.

Teachers and students need a safe, digital space that allows for student-to-student and teacher-to-student connection, communication, and collaboration. This can be accomplished if teachers use Microsoft Teams, Google Classroom, or a Learning Management System like Blackboard. When it comes to Microsoft Teams, teachers can create a standard (public) where all students in the class can view and interact with each other, the teacher, and the content. Alternatively, teachers can add specific students to private channels where they can connect, communicate, and collaborate. Standard channels can be for whole-class activities such as "Daily Check-Ins" while private channels can be for small group activities, tasks, and discussions. The following picture is a sample class in Microsoft Teams with both standard and private channels that foster SEL. Private channels have a padlock next to the numbered names in the channel list.

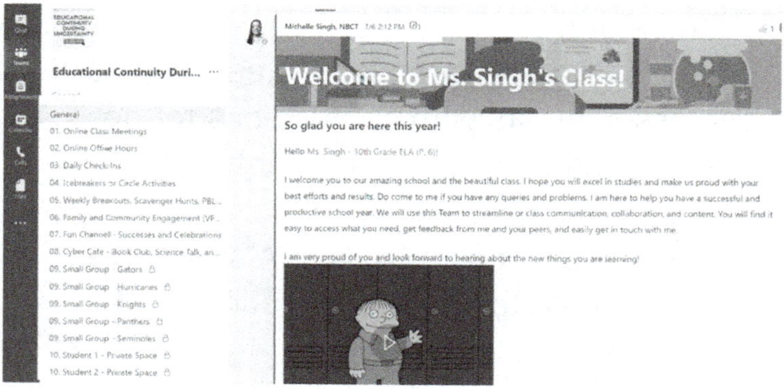

In addition, it is important that teachers continue to foster the CASEL Competencies in the online learning environment. CASEL is a framework that promotes SEL. Teachers can include these competencies in what students do and complete for learning tasks and assign:

1. Self-awareness (e.g., emotions)
2. Self-management (e.g., organizational skills)
3. Social awareness (e.g., respect for peers and others)
4. Relationship skills (e.g., teamwork)
5. Responsible decision-making (e.g., reflection)

Teachers can provide SEL opportunities for students that align with the CASEL SEL competencies. Here are some examples:

1. CASEL's Self-Awareness -- Mood Meter check-ins at the beginning of live lessons or online class meetings
2. CASEL's Self-Management -- Morning messages (secondary) or morning meetings (elementary)
3. CASEL's Social Awareness -- Learning tasks and assignments with connections to the feelings and emotions of characters or learning topics/scenarios
4. CASEL's Relationship Skills -- A space like a "Fun Channel" to share celebrations and successes
5. CASEL's Responsible Decision-Making -- Daily check-in prompts students can reflect upon and respond to in a safe space like a Microsoft Teams channel or Flipgrid topic

Self-Awareness | Mood Meter Check-Ins

At the beginning of live online class meetings, one of the first things teachers should do is check how their students are feeling. This can be done with a Mood Meter check-in. The Mood Meter was produced by the Yale Center for Emotional Intelligence's RULER program. It identifies feelings on different spectrums. The feelings are related to energy and pleasantness. Similar words for feelings are categorized by four colors on the Mood Meter. The sample below is a simple depiction of the Mood Meter.

Anxious	Content
Sleepy	Thankful

Teachers can ask students to identify a feeling from the Mood Meter. Students can then explain why that feeling applies to them. Teachers can also encourage students to track their mood throughout the day using the Mood Meter. This helps students build self-awareness.

Self-Management |Morning Messages or Meetings

Teachers can help students organize their day with morning messages and morning meetings. Elementary teachers can meet with their students in a virtual circle time for morning meetings. They can use Microsoft Teams, Zoom, Google Meet, or another virtual meeting platform.

Morning Meetings can include the following structure. These are adapted from *Morning Meetings* by Responsive Classroom:

1. Greetings and Hellos – teacher and students greet each other by name
2. Sharing and Feedback – teacher and students share and react to events and news
3. Lively Group Activity – teacher and students join in an activity together like a song or a Go-Noodle dance
4. Goal Setting Message – teacher sets the day up for the student with goals and "I can" statements

Secondary teachers can send recorded messages to help students start their day on the right track. Teachers can use Flipgrid Shorts, PowerPoint Recording, Microsoft Stream, Prezi Video, Loom, or Vocaroo to record videos and screencasts. Morning Messages can be a video or voice announcement for students. The announcement can include daily activities, clarification on any issues from previous days, procedures for learning tasks and assignments, tips for success, and an encouraging message at the end.

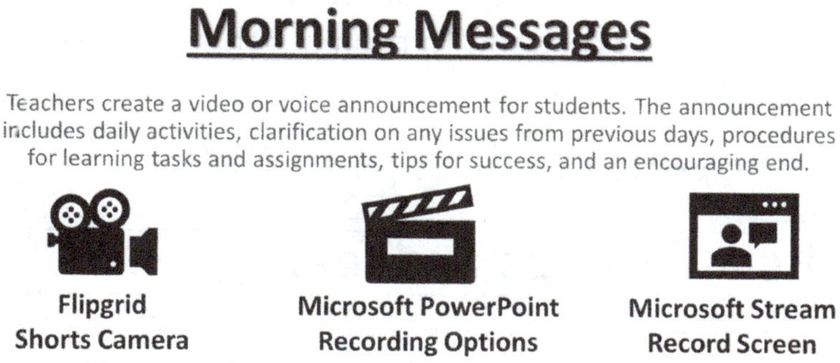

Social Awareness | Learning Connections

Teachers can help students build social awareness when they create learning tasks and assignments that have real-world connections. These activities can include some of the following components for learning connections to happen:

- Selecting reading materials and texts like current events, short stories, fiction, and non-fiction that align with the topics and curriculum
- Finding relationships to and recognizing the feelings and emotions of characters or people in the reading materials
- Connecting topics from the learning activities to the students' backgrounds and communities

- Allowing students to find parallels between themselves, their communities, and the lessons
- Activating students' background with a place for them to respond to prompts that tie into the lesson

Relevant reading materials

Relationships and feelings

Backgrounds and communities

Parallels

Prior knowledge

The example below is a post in a Microsoft Teams channel that aligns with social awareness and learning connections.

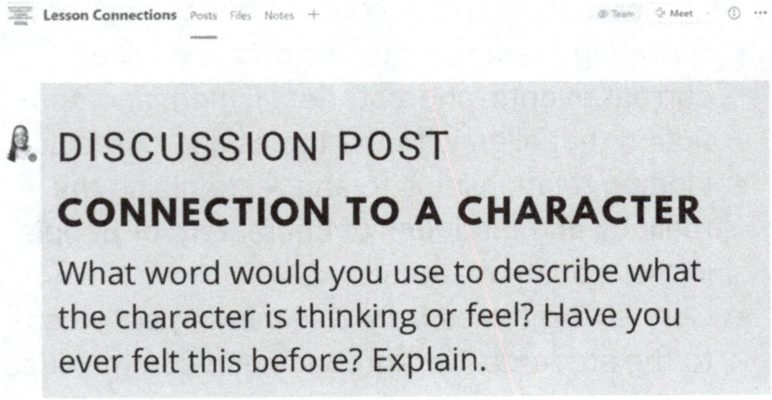

Relationship Skills | Microsoft Teams Fun Channel

Teachers can help students establish and maintain relationship skills by providing a space for them to communicate. In Microsoft Teams, teachers can create a "Fun Channel" for students to share celebrations and successes. They can learn to encourage one another by leaving praise and replies for each other in this channel.

Teachers can even have a "Parking Lot" channel where students can ask for help. Similar support topics can be created for students to respond to using Flipgrid, sharing, responding, and getting help.

See an example of praise in the sample Microsoft Class Team.

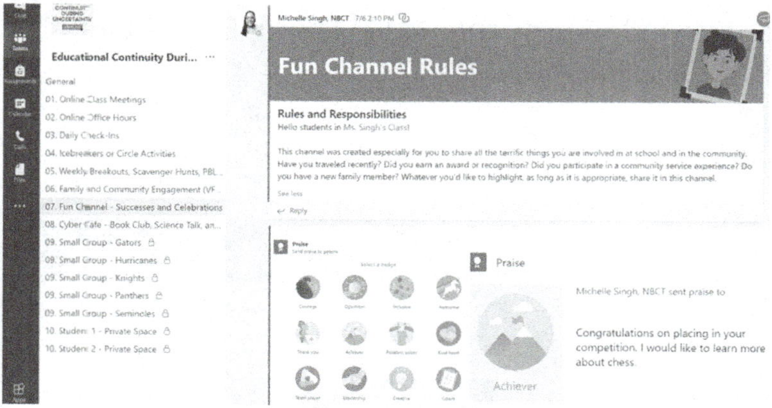

Responsible Decision-Making | Daily Check-In Prompts

Teachers can help students improve their responsibility and decision-making abilities. With Daily Check-In prompts (secondary) and goal sharing during morning meetings (elementary), students can reflect, plan, and evaluate their responsibilities for the day.

Teachers can also establish a predictable daily routine to guide students in executing their learning goals for the day. This information can be shared in the 'Daily Check-Ins' channel in Microsoft Teams or as a post in Google Classroom or a Learning Management System like Blackboard.

The following example of a Microsoft Teams daily check-in features the Rose, Thorn, Bud activity from *Mindful Schools*.

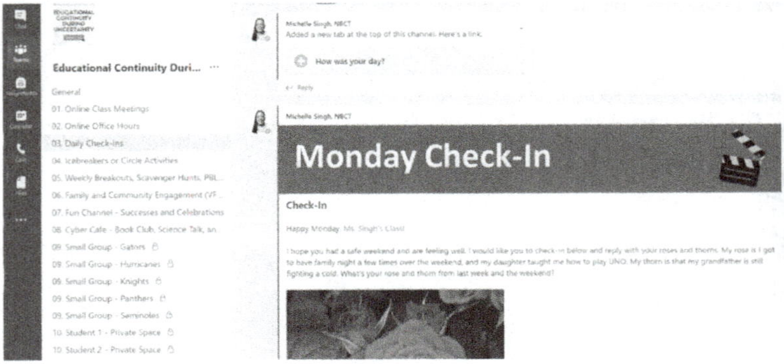

COGNITIVE CONTINUITY

Cognitive continuity relates to how teachers strategically help their students develop higher-order thinking skills (HOTS) and metacognition.

HOTS and metacognition foster critical thinking skills and creative abilities in students. When students are exposed to HOTS and metacognition, they become more aware of their learning and what they need to do to be successful learners. Students are also given more options to create content. They are no longer consumers of information in this process.
Teachers can help students become critical thinkers and creators of content when they:

- Incorporate learning activities that foster HOTS with problem-solving tasks, learning challenges, open-ended questions, and re-teaching of concepts learned. The infographic below provides information about HOTS. The information was adapted from *Literacy Strategies for Grades 4-12* by Karen Tankersley. Teachers can use the verbs to design learning tasks which students will complete or do.

Higher Order Thinking Skills

Synthesis → Create	• choose, combine, compose, construct, create, design, develop, formulate, hypothesize, invent, make, make up, originate, organize, plan, produce, role play
Evaluation → Design	• appraise, assess, check, compare, conclude, criticize, critique, defend, justify, support
Analysis → Analyzing	• analyze, categorize, classify, differentiate, distinguish, identify, infer, point out, select, subdivide, survey
Application → Apply	• apply, choose, dramatize, explain, generalize, judge, organize, paint, prepare, produce, select, show, sketch, solve, use
Comprehension → Understand	• classify, defend, demonstrate, distinguish, explain, express, extend, give an example, illustrate, indicate, interrelate, infer, judge, match, paraphrase, represent, restate, rewrite, select, show, summarize, tell, translate
Knowledge → Remember	• choose, describe, define, identify, label, list, locate, match, memorize, name, omit, recite, recognize, select, state

- Include metacognitive tasks that ask students to think about and monitor their thinking processes; reflect on their challenges and successes; develop an outlook of what success looks like; and deepen their understanding of a growth mindset. The stages of metacognition can be seen in the infographic below which was adapted from the Science of Learning series about Metacognition.

Metacognition

PLANNING	MONITORING	ASSESSMENT
Students think about what they know and want to know (their goals).	Students think about how they will learn during the process of learning.	Students think about and evaluate their understanding of the learning.

More specifically, teachers can integrate HOTS and metacognitive strategies for their students with these strategic task options:

1. Reflective questioning
2. Self-assessment
3. Digital portfolios
4. Student-created content
5. Authentic activities

Reflective Questioning

Reflection is challenging but necessary. It is something teachers must teach their students to do. With consistency, teachers can help students use this metacognitive strategy to connect their learning, extend their thinking, and breakthrough challenges. To do this, teachers can use a Microsoft or Google Form, Word or Google document, OneNote page, or Flipgrid video to have students reflect on the questions below for new topics/units. These questions were adapted from Beth Holland's article, *The Art of Reflection*, which was featured on Edutopia.

	CONNECT: How do the learned concepts/skills connected to students' prior knowledge?
	EXTEND: How did the students' learning extend their thinking (e.g., other subject areas)?
	CHALLENGE: What are still some challenges students face with the new concepts/skills?

Self-Assessment

Self-assessment is an important metacognitive strategy that can strengthen student's thinking. When students self-assess during their learning process, they improve their awareness and ability to acknowledge what works best for them. They become self-directed learners who can take control of learning situations.

Teachers can use these questions or prompts to help students self-assess. Students can respond using a Microsoft or Google Form, Word or Google document, OneNote page, or Flipgrid video the teacher sets up. The following metacognitive prompts for self-assessment are adapted from *What You Need to Know About Metacognition* by The Learning Agency Lab.

1. *What do I know about the topic/concept/skill already?*
2. *How do I know this information about the topic/concept/skill?*
3. *What new information did I learn about the topic/concept/skill?*
4. *How would I explain this new information about the topic/concept/skill to a friend?*
5. *Did learning this new information change my thinking?*
6. *Is there anything else I can learn about the topic to deepen my understanding?*

Digital Portfolios

Digital portfolios are a fantastic way for students to document their learning growth. Microsoft OneNote, Nearpod, and Wakelet are great tools for students to use for their digital portfolios. They give them options for multi-modal sharing through audio, video, text, and visuals. It also helps them organize their thinking based on assignments/artifacts for subjects, content, grading periods, or weekly entries. Teachers can consider these reflective questions for students to discuss in their digital portfolio entries for each week:

1. *What was your learning goal for this week? Write your "I can..." statement(s).*
2. *What did you learn this week?*
3. *How would you describe your process of learning the concept or skill? Did you use any strategies or resources to help learn?*
4. *Is there something you did or created that makes you proud? Select a piece of work you did this week and upload or include a copy of it.*
5. *What questions do you still have about the concept or skill?*
6. *Are there concepts or skills that are still unclear?*
7. *Did you experience any struggles with the learning?*
8. *How did you deal with these struggles? Did you use any strategies or resources?*
9. *Can you suggest another way to teach this week's concept or skill to your peers?*
10. *What are your learning goals for next week?*

See this sample Microsoft OneNote digital portfolio. The student in this example included an artifact of their learning along with responses to the reflective questions above. These questions allow the students to reflect on their learning for the week. Feedback from the teacher is also included in the example.

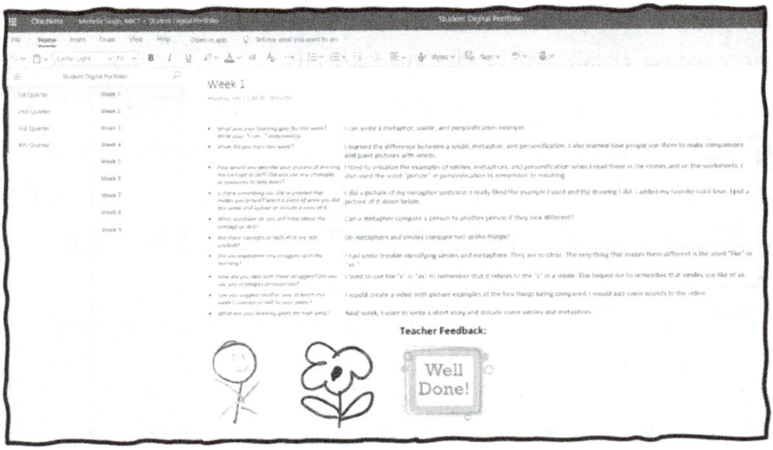

Student-Created Content

Teachers can help foster the growth of HOTS for students when they allow them to be creators of the content they learn. When students create, they use their highest thinking skills to demonstrate knowledge, showcase expertise, and highlight learning style strengths. See some examples of what students can create in the chart. These examples are aligned to various learning styles. They also provide a digital product option for students to create for the learning demonstration:

Visual	***Digital Game/Comic*** Design a 5-part game or 5-page comic with colorful, creative, images and text related to the topic.
Verbal	***Podcast*** Produce a 5-minute podcast. The featured discussion should include information about the topic and related articles. Guest speakers may be included.
Read/Write	***Webpage/Blog*** Build a webpage or blog. The webpage or blog should include 5-sections or entries with text, images, infographics, and other media related to the topic.
Auditory	***Musical Playlist*** Construct a musical playlist with 5-songs that connect to the topic. Have a written commentary connection to accompany each song.
Kinesthetic	***Digital Story/Vlog*** Make a 5-minute digital story or 5-video vlog related to the content. Include commentary on the topic and related current events.

Authentic Activities

Authentic activities expose students to HOTS, metacognition, and 21st-century skills. Authentic activities also capture students' attention and make learning relevant. They require students to plan, create, collaborate, communicate, reflect, and demonstrate knowledge of content/standards. Some examples of authentic activities include:

1. Research, evaluation of sources, and synthesis
2. Character, relationship, and/or situational analysis
3. Reflective writing (e.g., Double-Entry Journals, Reading Logs)
4. Peer discussion activities (e.g., Think-Pair-Share, SCAMPER)
5. Group learning tasks (e.g., Jigsaw, Six Thinking Hats)
6. Peer critique and evaluation using rubrics

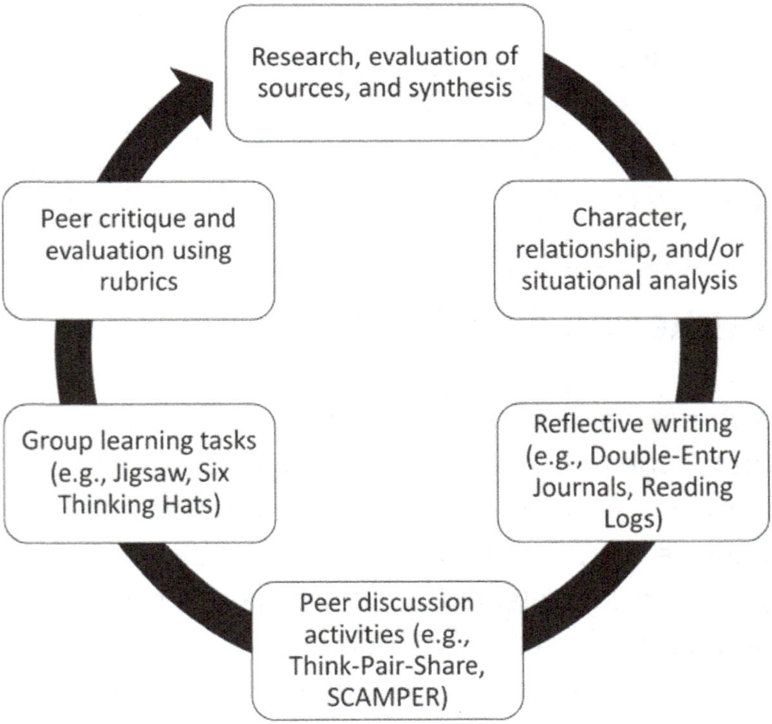

ACADEMIC CONTINUITY

Academic continuity relates to the content taught, the delivery of the content, and the assessment of the content.

During online learning, teachers must place compassion over content. In this way, they support students and families while meeting learning goals with shorter, simplified, and more targeted lessons. Teachers can also have flexibility in deadlines, content, delivery, and assessments.

Most importantly, teachers can provide a predictable place and platform for students to access directions, meetings, announcements, discussions, assignments, etc. It is necessary for students to have an organized, user-friendly, and easily accessible structure for their learning. With structured guidelines like the example below which was adapted from *4 Steps for Powerful Distance-learning Experiences* by Douglas Fisher and Nancy Frey, students will have clear expectations for their contributions to the class and learning.

Attend	What are the dates and times of the live, online class meeting(s) this week?
Read	What readings and readings tasks must be completed by students this week?
View	What videos can students watch to help with the concept/skill for the week?
Listen	Are there audio recordings or messages students must listen to for help?
Discuss	What discussion questions must students respond to with their classmates?
Solve	What collaborative activities and tasks must small groups work together to solve?
Practice	What individual assignments must students complete and submit for the week?

Furthermore, teachers can create purposeful learning activities for students that achieve academic continuity. Considerations for learning activities include:

1. Interactive, direct instruction
2. Ongoing instruction and support
3. Personalized learning
4. Gamification
5. Research-based instructional strategies

Interactive, Direct Instruction

Teachers can use a virtual meeting platform like Zoom, Google Meet, and Microsoft Teams meetings for synchronous, online class meetings. The online meeting platform should have capabilities for chat; content sharing for applications like Microsoft PowerPoint, websites like Nearpod or Google Slides, and the desktop; and breakout rooms for students to work in small groups or to have a place to work individually for silent-work time.

Teachers can take advantage of features like the ones just mentioned to make the online meeting experience interactive, thoughtful, purposeful, engaging, and fun for students. The chart below includes recommendations of what the teacher and students can do from the start to finish during a live online class meeting.

STUDENT ACTIONS	TEACHER TASKS	EXAMPLES OF TASKS FOR STUDENTS TO COMPLETE
Participate Speak	Icebreaker	Quotes, fun facts, song/dance/play (e.g. Go Noodle, Yoga Kids)
Learn Consume	Content	Videos, slide shows, reading/text demonstrations, whiteboard
Write Reply	Chat	Open-ended questions, reflection questions, feedback
Summarize Synthesize	Notes	Cornell notes, collaborative notes, annotations (using Kami app)
Collaborate Create	Breakout	Scavenger hunts, projects, investigations, games/simulations
Practice Perform	Assessment	Exit tickets, polls (using Microsoft Forms, Kahoot, Quizizz, Quizlet Live)

Ongoing Instruction & Support

Teachers can offer ongoing instruction and

support to students with pre-recorded video lessons. These video lessons can help teachers determine what students know about a topic or skill. This in turn will be a guide for teachers as they plan the next steps for students.

When creating or curating video lessons, teachers can create learning experiences instead of viewing experiences. They can include activities for students to complete before, during, and after the video lessons. Tools like Flipgrid, Microsoft Stream (add a survey, poll, or quiz), PowerPoint (upload to Stream), Edpuzzle, and Nearpod interactive video can be used.

Before recording video lessons, teachers can prepare the content they will teach. It also helps to have a brief script or outline to map out what they will say.

Furthermore, teachers can use these suggestions as they record their video lessons:

- Be simple and clear
- Slow down the speech
- Use an upbeat, cheerful tone
- Add graphics/animations/visuals
- Keep the time between 2 & 5 minutes

Finally, teachers can add engagement and interactivity with these options which were adapted from Flipped Learning Flow for Blended or *Online Classes* by Catlin Tucker:

Before the Video Lesson	During the Video Lesson	After the Video Lesson
• Students reply to a prompt or question about the topic	• Students reply to open-ended questions	• Students fill out an exit ticket
• Students brainstorm what they know about the topic	• Students respond to true/false questions	• Students complete a simple quiz
• Students make predictions about the topic/learning	• Students complete multiple-choice questions	• Students respond to a writing prompt
• Students ask informed questions about the topic	• Students answer questions for a poll	• Students create notes or a concept map

Personalized Learning

Teachers can personalize learning for students to address their unique and diverse needs, interests, strengths, skills, and home situations. They can do this by making sure the information, lessons, learning activities, and content they share are easily accessible and multimodal. This can be achieved with the Microsoft accessibility options like Microsoft Translator, Dictation, and Immersive Reader. These tools are embedded in Microsoft 365 applications like Word, PowerPoint, Excel, OneNote, Forms, and Flipgrid. They are also available on platforms like Wakelet, Nearpod, and Buncee. These tools can help teachers differentiate their content and the process of teaching the content.

Teachers can also personalize learning for students. They can do this by providing voice and choice options for students to demonstrate their learning. Teachers should differentiate the products for students to create with tools like Flipgrid, Sway, PowerPoint, Nearpod, Wakelet, Buncee, and Microsoft OneNote. Students can then showcase their knowledge, skills, and strengths in a variety of ways.

Gamification

Teachers can increase collaboration, competition, and enthusiasm in their lessons (synchronous or asynchronous) with gamification. Gamification involves adding gaming elements to the online learning environment to foster play, competition, achievement, creativity, and collaboration. Recognizing student achievement is a necessary piece of the gamification puzzle, as well. Teachers can create personalized badges for students using Credly, a digital badging website. Teachers can consider these applications and programs for classroom gamification (some are free, and some require payments or subscriptions):

- Gimkit
- Kahoot
- Microsoft Make Code
- Minecraft Education Edition
- Nearpod's Time to Climb
- Quizalize
- Quizizz
- Quizlet Live
- Scratch
- Socrative

Research-Based Instructional Strategies

Teachers can use these research-based instructional strategies to achieve academic continuity in a virtual setting with minor adjustments. Examples like these can be used with students with great results:

- **Jigsaw** -- The teacher uses breakout rooms during meetings. Students meet in their expert groups to discuss the content shared by the teacher. Each expert group prepares a mini-lesson to teach the whole class (home group) about the topic.
- **Reciprocal Teaching** -- The teacher brings student-experts to co-teach with them during synchronous online class meetings. Student-experts can re-teach and summarize concepts in the lesson. Student-experts can also create or add to video lessons shared with the class.

- **Think-Pair-Share (Ink-Pair-Write) --** The teacher poses the question and asks students to think about it individually and or ink their thoughts on Whiteboard collaboratively. They then send students to breakout rooms to pair and discuss. Students come back to write/share in the chat.
- **Gallery Walks** -- The teacher collects student digital products and adds them to a Padlet. The teacher shares the Padlet link with the class and asks students to leave comments. The teacher can also show Padlet and have students present during a class meeting.
- **Cornell Notetaking** -- The teacher provides guided notes for students using the Kami app, Word Online, etc. During meetings, students can work in breakouts to complete the notes or they can work individually during silent notetaking in a meeting.

BIBLIOGRAPHY

Cox, Janelle. "Teaching Strategies That Enhance Higher-Order Thinking." TeachHUB, K-12 Teachers Alliance, 16 Oct. 2019, www.teachhub.com/teaching-strategies/2019/10/teaching-strategies-that-enhance-higher-order-thinking/.

Fegely, Cherner A. "Digitalizing Gallery Walks: A Method for Student-Centered Feedback and Engagement." Teaching Online Pedagogical Repository, University of Central Florida Center for Distributed Learning, 2017, https://topr.online.ucf.edu/digitalizing-gallery-walks-method-student-centered-feedback-engagement/.

Fisher, Douglas, and Nancy Frey. "4 Steps for Powerful Distance-Learning Experiences." SmartBrief, 16 July 2020, https://www.smartbrief.com/original/2020/07/4-steps-powerful-distance-learning-experiences.

Fisher, Douglas, and Nancy Frey. "No Instructional Minute Wasted." No Instructional Minute Wasted - Educational Leadership, ASCD, 1 June 2020, www.ascd.org/publications/educational-leadership/jun20/vol77/num09/No-Instructional-Minute-Wasted.aspx.

Gonzalez, Jennifer. "9 Ways Online Teaching Should Be Different from Face-to-Face." Cult of Pedagogy, 5 July 2020, www.cultofpedagogy.com/9-ways-online-teaching/.

Holland, Beth. "The Art of Reflection." Edutopia, George Lucas Educational Foundation, 17 Dec. 2017, www.edutopia.org/blog/digital-portfolios-art-of-reflection-beth-holland.

Johnson, Ben. "How to Coach Parents Who Are Teaching at Home." Edutopia, George Lucas Educational Foundation, 8 July 2020, www.edutopia.org/article/how-coach-parents-who-are-teaching-home.

"Metacognition | WHAT YOU NEED TO KNOW ABOUT METACOGNITION." The Learning Agency Lab, 2020, www.the-learning-agency-lab.com/science-of-learning-research-metacognition.html.

Pitts, Christine. "3 Ways to Support Students' Emotional Well-Being During the Pandemic." Edutopia, George Lucas Educational Foundation, 22 May 2020, www.edutopia.org/article/3-ways-support-students-emotional-well-being-during-pandemic.

Poth, Rachelle Dene. "Metacognition and Why It Matters in Education." Getting Smart, 14 Oct. 2019, www.gettingsmart.com/2019/10/metacognition-and-why-it-matters-in-education/.

Taylor, Janet. "How to Give Authentic Assessment for Online Learning." The Art of Education University, 3 June 2020, https://theartofeducation.edu/2020/06/05/how-to-give-authentic-assessment-for-online-learning/.

Tucker, Catlin. "4 Strategies Designed to Drive Metacognitive Thinking." Catlintucker.com, 13 June 2019, https://catlintucker.com/2018/09/4-strategies-metacognitive-thinking/.

Tucker, Catlin. "8 Ideas Designed to Engage Students in Active Learning Online." Catlintucker.com, 9 July 2020, https://catlintucker.com/2020/07/8-ideas-designed-to-engage-students-online/.

Visit **www.lctelearning.com**
to view a Resource Page with
links to the technology tools,
teaching strategies, and the
resources discussed in this book.